Recycling Organic Waste: From Urban Pollutant to Farm Resource

GARY GARDNER

Jane A. Peterson, *Editor*

WORLDWATCH PAPER 135
August 1997

THE WORLDWATCH INSTITUTE is an independent, nonprofit environmental research organization in Washington, DC. Its mission is to foster a sustainable society in which human needs are met in ways that do not threaten the health of the natural environment or future generations. To this end, the Institute conducts interdisciplinary research on emerging global issues, the results of which are published and disseminated to decisionmakers and the media.

FINANCIAL SUPPORT for the Institute is provided by the Nathan Cummings Foundation, the Geraldine R. Dodge Foundation, The Ford Foundation, the Foundation for Ecology and Development, The William and Flora Hewlett Foundation, W. Alton Jones Foundation, John D. and Catherine T. MacArthur Foundation, Charles Stewart Mott Foundation, The Curtis and Edith Munson Foundation, The Pew Charitable Trusts, Rasmussen Foundation, Rockefeller Brothers Fund, Rockefeller Financial Services, Summit Foundation, Turner Foundation, U.N. Population Fund, Wallace Global Fund, Weeden Foundation, and the Winslow Foundation.

THE WORLDWATCH PAPERS provide in-depth, quantitative and qualitative analysis of the major issues affecting prospects for a sustainable society. The Papers are written by members of the Worldwatch Institute research staff and reviewed by experts in the field. Published in five languages, they have been used as concise and authoritative references by governments, nongovernmental organizations, and educational institutions worldwide. For a partial list of available Papers, see back pages.

Table of Contents

Tables and Figures

The views expressed are those of the author and do not necessarily
represent those of the Worldwatch Institute; of its directors, officers,
or staff; or of its funding organizations.

ACKNOWLEDGMENTS: I would like to thank Sidonie Chiapetta, Laurie Drinkwater, Robert Goodland, Josefina Mena Abraham, Laura Orlando, Mark Ritchie, David Wedin, Ray Weil, and my colleagues Lester Brown, Chris Flavin, Jennifer Mitchell, and Molly O'Meara for their helpful comments on an early draft of this paper. Worldwatch interns Yasmin Daikh and Giovanna Dore provided thorough research assistance. Jane Peterson, our editor, and Jim Perry, Denise Byers Thomma, and Mary Caron of our communication staff were instrumental in sharpening the paper's message. Liz Doherty was quick and accurate in the production process, in spite of a tight schedule. And Sally Bolger provided unfailing good humor and unwavering moral support. To all, a heartfelt thank you.

I am grateful to the Wallace Genetic Foundation for its generous financial support of this project.

GARY GARDNER is a Research Associate at the Worldwatch Institute, where he writes on agriculture and water issues. Since joining the Institute in 1994, he has written chapters in *State of the World* and contributed to *Vital Signs* and *World Watch* magazine. He also wrote Worldwatch Paper 131, *Shrinking Fields: Cropland Loss in a World of Eight Billion*, released in July 1996.

Mr. Gardner was previously a project manager at the Soviet Nonproliferation Project, a research and training program run by the Monterey Institute of International Studies in California. While there, he authored *Nuclear Nonproliferation: A Primer*. Mr. Gardner spent two years helping Peruvian women's groups develop urban small livestock projects. He holds master's degrees in Politics from Brandeis University, and in Public Administration from the Monterey Institute of International Studies. He received his bachelor's degree from Santa Clara University.

Introduction

In 1876, a German chemist studying the agricultural history of North Africa became increasingly troubled over the fate of that region and its implications for his day. In the first century AD, North Africa's fertile fields were supplying two thirds of the grain consumed in Rome. But the nutrients and organic matter in that food were not returned to the farms where they originated; instead, they were flushed into the Mediterranean. By the middle of the third century, the one-way flow of nutrients out of North Africa's grainland soils, along with declining levels of organic matter, had contributed to the region's tumble into environmental and economic decline.[1]

The chemist, Justus von Liebig, worried that Europe's rapidly expanding cities also depended too heavily on one-way nutrient flows, with consequences that would eventually undermine both urban and agricultural areas. To solve the problem, he invented chemical fertilizer, essentially a mixture of condensed and easily transportable nutrients that made it possible to escape dependence on recycling organic matter. The new fertilizer revived the fertility of nutrient-depleted farmland. And because a ton of this plant food could pack as many nutrients as dozens of tons of organic matter, it could be shipped cheaply over great distances. Cities could now expand, and food could be imported from great distances, without concern for returning urban garbage and sewage to farmlands. Thus, garbage and sewage became waste products to be discarded, rather than soil builders to be reused.

Today, nearly 3 billion of us—half of the human family—live in cities, more dependent than ever on long, one-way flows of nutrients and organic matter. But reliance on linear flows instead of the traditional organic "loop" comes at a price that is paid at both ends. To start with, many regions of the globe are now overfertilized, a trend with consequences well beyond the farm. Drinking water in several European countries is contaminated with fertilizer runoff. Species diversity is reduced in some land-based ecosystems by excess applications of nitrogen. The quality of organic matter declines, and plant diseases become more prevalent, in soils dependent on manufactured fertilizer. And aquatic life in rivers, lakes, and bays suffocates as blooms of algae fatten on nitrogen and phosphorus that have leached and eroded from these soils. In short, a host of new problems arise once the circular flow of nutrients (essential for plant growth) and organic matter (essential for soil health) is disrupted and made into a linear flow.[2]

At the disposal end of the linear flow, meanwhile, natural sources of nutrients and organic matter in urban garbage and human excreta are increasingly difficult to eliminate safely. Landfills for solid waste are not only near capacity in many countries, they are leaking toxic chemicals into groundwater and methane into the atmosphere. Discarded garbage piles high on street corners in many developing countries, spawning rats and disease. And human waste is either dumped indiscriminately or mixed with industrial chemicals in urban sewers, which makes safe disposal much more difficult. In any case, sewage systems are expensive and water intensive—flush toilets account for 20-40 percent of residential water use in sewered cities of developed countries—making them an inaccessible luxury for the growing number of cash-strapped and water-short cities in the developing world.[3]

Returning nutrients in organic matter to farm soils—"closing the organic loop"—would help alleviate all of these problems. Urban wastes such as food scraps, paper, and yard clippings can be composted and applied to soils, thereby improving soil structure, supplying nutrients, and suppressing disease. Indeed, nutrients in the garbage and yard wastes of states belonging to the Organisation for Economic Co-operation and Development (OECD) equal some 7 percent of the nutrients in fertilizer applied in those countries, and the level is much higher in many developing countries. In addition, nutrients in discarded human waste in OECD countries equal another 8 percent of applied fertilizer. While urban organic wastes will not displace fertilizer entirely, they can help reduce excessive fertilizer use (and the pollution this causes) as they build healthier soils.[4]

Recycling organic matter would also ease the pressure on costly waste disposal facilities. Organic matter accounts for a third of inflows to landfills in industrialized countries, and as much as two thirds in developing countries, and is largely to blame for the acidic leaching and methane problems that these facilities generate. Meanwhile, opting for a "dry" system of human waste management— through the use of composting toilets, for example—would free up clean water for more vital uses, and avoid costly infrastructure construction as well.[5]

Closing the organic loop would help alleviate many urban problems.

Before extensive reuse of organic material can take place, however, certain changes in agricultural production and trade practices must occur. As organic flows extend across oceans, and as agricultural production becomes more specialized and intensified, nutrients inevitably accumulate in some areas. Centralized livestock facilities, for example, like the giant poultry- and hog-raising operations in the United States, buy feed from far away, and then have trouble disposing of all the manure they produce. Manure is one nutrient source that has commonly been recycled, for livestock and crops located on the same farm easily fed each other. But as livestock operations increase in size and become separated from agriculture, more and more of this resource is viewed as waste material.[6]

Such regression is also evident in some developing countries that mimic the nutrient management practices of industrialized nations. China, for example, used organic sources for more than 98 percent of the fertilizer applied to soils in 1949; today, because of rising labor costs, the share is less than 38 percent. On the other hand, some industrialized regions are paying greater attention to reuse of organic matter as the problems created by linear flows of nutrients mount. In the U.S., 23 states now restrict the inflow of grass clippings to landfills; this material is composted or re-used as mulch. And well over a third of U.S. and European sewage sludge is now applied to land, though often with only minimal precautions for safe reuse.[7]

Continued progress in recycling organic material requires that it be viewed as a natural resource, not as waste matter. Such a shift in perspective will require education on many levels. Policymakers and citizens will need to learn to manage organic matter in ways that facilitate its reuse. Processors of organic matter, such as compost makers, will need to tailor their products to the diverse needs of different soils and crops. And farmers will need to understand how organic matter works in soils, and how they can avoid overuse of chemical fertilizers. Once this educational process is complete, other steps will follow naturally. Communities will close dumping sites to organic materials as people adopt environmentally supportive disposal technologies and management practices—such as garbage and sanitation systems that segregate organic matter from harmful chemicals and non-organic wastes. Together, these steps will promote circulation of more organic matter.

Recycling organic wastes and returning them to productive soils would be a large step toward sustainability for the world's cities and national economies. But the current trend in most of the world—toward greater dependence on extended, one-way nutrient flows facilitated by heavy fertilizer use—promises increased ecosystem disruption, greater waste disposal problems, and eventually a negative effect on food production itself. As policymakers grapple with the multiple problems of today's burgeoning cities, they would do well to ponder the multiple advantages that emerge from the wise reuse of organic matter. By retapping this important natural resource, decisionmakers can ease the urban burden on several fronts.

The Cost of Breaking the Loop

When the natural circular flow of organic material is broken, two challenges immediately arise: the flow must be fed at one end, and emptied at the other. What once occurred automatically in a cycling system, where feed and waste chased each other perpetually, now requires conscious intervention at either end. The inflow challenge is typically met with a steady stream of manufactured fertilizer, while disposal is handled in several ways, depending on the material's final form: sewage, garbage, or manure. These endpoint manipulations make linear flows possible. But they also create new problems. Today, the price for breaking organic loops is a diverse set of problems, from pollution and poor soil health caused by excessive dependence on fertilizer, to difficulty disposing of nutrient-laden wastes cleanly.

At the front end of the organic pipeline is a set of problems created largely by the overuse of fertilizer, the pipeline's "pump." When it was invented, fertilizer was viewed as a godsend: by separating the major nutrients from their host environments—nitrogen from the air, and phosphorus and potassium from rocks and minerals—scientists developed a potent and portable resource that eliminated the need to recycle bulky organic matter. Fertilizer also increased crop yields, and in combination with cheap transportation it allowed the development of large cities, which could grow without concern for returning organic wastes to the ever-more-distant fields on which they depended for food. Unforeseen, however, was the growing human and environmental toll that would result from excessive dependence on chemical fertilizer, a toll now felt even at the global level.

Fertilizer production has spurred a sharp increase in the global rate of nitrogen fixation—the process that converts nitrogen to a form usable by many living organisms. Nitrogen is now fixed at more than twice the natural, pre-industrial rate, which essentially means a boost in fertility over most of the planet. (See Figure 1.) This surge is caused by a variety of human activities, chief among them being fertilizer production, which has grown more than ninefold since 1950. Because half of the manufactured fertilizer used in human history has been applied only since 1982, the greatest surge in nitrogen (and phosphorus and potassium) levels is quite recent, and its full effects are yet to be understood.[8]

Many of the consequences of the planet's overfertilization are more pernicious than might be expected. The presence of fixed nitrogen at greater than natural levels, for example, has been shown to reduce plant diversity in prairie ecosystems at an alarming pace. In a recent 12-year study, scientists applied nitrogen to 162 test plots of Minnesota grasslands at varying rates. The nitrogen spurred the growth of plants that were best able to take it up—but at the expense of plants that were less well adapted. Indeed, species diversity declined by more than 50 percent. This loss of diversity is consistent with the experience of parts of northern Europe, where high levels of nitrogen deposition have converted species-rich heathlands to species-poor grasslands.[9]

The loss of species diversity, lamentable in itself, also made the ecosystem "leakier," and therefore more polluting. Because the invasive species were less able to store nitrogen than the native grasses they replaced, nitrogen leaching—an important source of water pollution—increased over the study period as the ecosystem became biologically impoverished. The Minnesota study is another contribution to the growing body of research documenting the harmful impact of excessive levels of nitrogen, once considered a relatively benign nutrient.[10]

FIGURE 1

Natural and Human Sources of Nitrogen Fixing

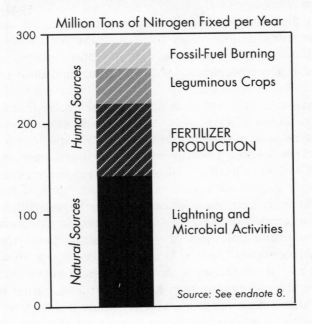

Million Tons of Nitrogen Fixed per Year

Human Sources

Natural Sources

Fossil-Fuel Burning

Leguminous Crops

FERTILIZER PRODUCTION

Lightning and Microbial Activities

Source: See endnote 8.

Nutrient leaching can be especially high on a farm (a cropped ecosystem with little species diversity), particularly when manufactured fertilizer is used. A recently completed 15-year study by the Rodale Institute compared nitrogen budgets in three farming systems: one using manufactured fertilizer, one using manure, and the third using leguminous crops as nitrogen sources. The conventional fields leached 270 kilos of nitrogen per hectare, compared with 180 kilos on the manure-fed land, and only 110 kilos on the legume-cropped fields. Moreover, the conventional fields received relatively heavy inputs of nitrogen (a common occurrence on today's conventional farms), much more than was taken up by crops. This combination of high inputs and high leakage—akin to opening a faucet full-force into a sieve—meant that the conventional fields lost nitrogen in large quantities.

Indeed, after 15 years, soil nitrogen in the conventional
fields had decreased by 11 percent, while the manure-fed
fields gained nitrogen (which the soils stored for future use
by crops), and the legume-fed soils kept it roughly in balance. The study demonstrates the "leakiness" of conventionally fertilized soils, and the much greater capacity of
organic nitrogen amendments to increase or maintain soil
fertility.[11]

Nutrient leakage like that documented in the Rodale
experiment contributes heavily to water pollution. In an
aquatic equivalent of the Minnesota grasslands species losses, eroded or leached phosphorus and nitrogen promote
overgrowth of algae in rivers, lakes, and bays at the expense
of other species, including various fish. In fact, leached and
eroded nutrients help make agriculture the largest diffuse
source of water pollution in the United States. So extensive
is the agricultural pollution of the Mississippi River—the
main drainage conduit for the U.S. Corn Belt—that a "dead
zone" the size of New Jersey forms each summer in the Gulf
of Mexico, the river's terminus. Rich in fertilizer nutrients
that feed algae, the once productive area now has far fewer
fish and shrimp, which cannot compete with the decom-
posing algae for oxygen. The phenomenon is repeated on a
smaller scale around the world in countless rivers and lakes
that receive agricultural pollutants.[12]

Pollution caused by overuse of nitrogen and phosphorus is also harmful to human health. Nitrates in drinking
water can be converted to potential carcinogens when
digested by humans, and can cause brain damage or even
death in infants by affecting the oxygen-carrying capacity of
the blood. The OECD lists nitrate pollution as one of the
most serious water quality problems in Europe and North
America. Indeed, every member state of the European
Union has areas that regularly exceed maximum allowable
levels of nitrates in drinking water. The problem is expected
to worsen in developing countries whose fertilizer use is
accelerating, such as India and Brazil.[13]

All of these front-end problems could be ameliorated

to some degree if more organic wastes were recycled. Organic wastes typically contain the major nutrients supplied by fertilizer—nitrogen, phosphorus, and potassium—as well as a series of trace nutrients. And unlike fertilizer, these wastes contain organic material, which builds soil structure and creates a hospitable environment for plant roots that nurtures crop growth. At the national level, recycled organic material could supplant only a portion of total fertilizer use, because too little organic matter exists close enough to farms to provide all the nutrients needed by high-yielding varieties. But in combination with more efficient fertilizer use, organic recycling to cropland can reduce a major source of water pollution and ecosystem degradation in countries that use fertilizer heavily.[14]

In some regions, however, a shortage rather than a surplus of nutrients plagues agricultural soils. Poor farmers in many African countries, unable to afford enough fertilizer, essentially mine their soils, with more nutrients leaving for cities or other countries than are returned in fertilizer or organic matter. In the worst cases, nutrients leave agricultural soils three to four times faster than they are replaced. In Sub-Saharan Africa overall, fertilizer usage is so low that it replaces only 28 percent of the nitrogen, 36 percent of the phosphate, and 15 percent of the potash absorbed by crops. Given this clearly unsustainable situation, the region would benefit from greater cycling of organic material from cities to farming areas. Systematic use of what is now considered waste material could help to keep fertilizer use on these farms from reaching the excessive levels found in many industrial countries.[15]

At the back end of the organic pipeline is a different series of problems, most of which are related to waste disposal. Landfills in many industrialized countries, for example, are closing at a record clip. In the United States, the 8,000 landfills in operation in 1988 had dwindled to 3,091 by 1996, as many sites were unable to comply with federal environmental regulations, and as others simply filled up. While total capacity has actually increased in this decade (because landfills are now bigger) some areas are feeling a waste capacity squeeze. For example, New York City's Fresh Kills dump—the city's last remaining landfill, and the largest in the world, covering 1,200 hectares—is set to close in 2001. City officials are drawing up plans to export their garbage, some 13,000 tons *per day*, to other states. Other disposal options such as incineration and ocean dumping are expensive or environmentally problematic, or are banned outright.[16]

Organic material forms the bulk of the growing mountains of municipal waste: 36 percent of the waste flow in OECD member states is food or garden wastes. In developing countries, organic matter typically accounts for more than half, and often more than two thirds, of the total waste stream. Besides taking up space, rotting organic material pollutes land, water, and air by leaching acids and emitting methane, a greenhouse gas associated with climate change. New York's Fresh Kills dump emits more than 5 tons of methane and millions of gallons of acidic liquids each day; sanitation officials estimate that methane will continue to leak from the facility for 30 years after it is closed.[17]

The environmental costs and space needs of organic waste have raised official interest in reducing the tidal wave of trash into landfills. Several U.S. states have ordered inflows to dumps cut in half by the year 2000. In the U.K., authorities are working to reach a 25 percent recycling level for household refuse by the same year. And packed landfills in the Tokyo area have led the city to ponder a garbage collection fee to discourage waste generation. Composting organic matter, on the other hand, would free up space and reduce the pollution hazards created by decaying organic material. Indeed, the state of California sees composting as the natural solution to burgeoning dumps. But the challenge is great: the state will have to compost some 70 percent of urban organic wastes by the end of the decade to meet its waste reduction goals, a hefty boost from the current rate of 40 percent (which already represents an enormous increase from recycling levels of a decade ago).[18]

Human excrement is another resource-turned-waste product whose disposal is increasingly difficult as urban-rural organic loops are broken. For millennia, many cultures returned human waste to soils, and a few still do today. But increasingly the material is sewered, a disposal option that typically leads directly to landfills, incinerators, or oceans, dumping areas that are limited today or are easily polluted. The human toll from improper disposal (and from an unclean water supply, often a related problem) is intolerably high: some 2 million children die each year and billions of people become sick because of inadequate water and sanitation facilities. Yet the logical and traditional alternative—the recycling of sewage to farmland—is often unsafe because of the toxic industrial wastes that are mixed into many sewage flows.[19]

Organic material forms the bulk of the growing mountains of municipal waste.

Disposing of waste by sewer is also water intensive and expensive. But sewers remain the disposal option of choice, despite growing water scarcity in more and more regions. The United Nations' Comprehensive Freshwater Assessment, released in April 1997, notes that a third of the world's population lives in countries with moderate to high water stress; that share could reach two thirds by 2025. As levels of stress increase, the water needs of farmers, businesses, and households are unlikely to be met fully. Using dry methods of human waste disposal, such as composting toilets, would save a meaningful share of domestic water. These alternative sanitation technologies would also ease the strain on city budgets, since on-site systems cost only a fraction as much as sewer infrastructure. And local containment of human waste would increase the prospects for returning nutrients and organic matter to farm soils, not to mention the benefits for the health of the rivers and bays that formerly received them.[20]

That organic recycling could have so many diverse

benefits is not surprising; the problems cited here were created by the move away from a circular economy in the first
place. Therein lies the good news: just as straight-line organic flows have produced multiple problems, the return to
greater cycling of organic material promises a wide range of
benefits. The challenge is to send the organic portion of economic activity back to its source, as was done in earlier
times.

Organic Material Flows

If the flow of nutrients from farm soils were mapped, the
picture would resemble a tree, with a major trunk line
branching out to smaller flows as nutrients travel farther
from the farm. The nitrogen, phosphorus, and potassium in
Iowa soils, for example, might be taken up by corn, which
after harvest becomes food or feed. Each of these products,
in turn, branches out to one or more waste flows—sewage,
garbage, or manure. These nutrient-laden wastes then make
their way to thousands of landfills, incinerators, rivers, or
bays, which may be hundreds or even thousands of kilometers from the original soils. As nutrient flows multiply and
extend, the potential for returning nutrients to productive
soils diminishes.

This general picture, however, varies by country. Rural
economies have relatively simple and short nutrient flows—
corn may be consumed only locally and only as food, for
example—so returning nutrients to farm soils is relatively
uncomplicated, though often unpracticed. Industrialized
and urbanized economies face greater challenges in recycling, because their nutrient paths are long and multi-
pronged. Like a tree whose shape mirrors its root structure,
nutrient flows tend to reflect the complexity of the underly-
ing economy.

Crop nutrients are analogous to vitamins for humans.
They assist the fundamental process of photosynthesis—the

plant's use of light energy to transform carbon dioxide and water into organic compounds that give the plant its energy. Nutrients are found naturally in soils, but nitrogen, phosphorus, and potassium—the nutrients needed in major quantities for healthy plant growth—can also be added. This is done by applying organic matter and minerals, or by spreading manufactured fertilizer, the customary practice in conventional farming. Manufactured fertilizer is essentially a collection of nutrients drawn from natural sources and processed for use on plants. Nitrogen, for example, is taken from the atmosphere and "fixed" (converted to a form that plants can use) through an energy-intensive process. Phosphorus and potassium are mined, then processed into a form that is effective for use with crops.[21]

In addition to leaving soils through harvested crops, nutrients also erode away with wind and water, or leach down to an aquifer or out to a river or lake, or volatilize in a process akin to evaporation, changing to a gaseous form. Most of this paper focuses on nutrients that leave through harvested crops. But some of the efforts to return crop nutrients to farm soils—through composting of food wastes, for example—have the added advantage of reducing erosion and leaching as well.

Accounting for all nutrient flows of all crops in all

TABLE 1

Share of Nutrients from Domestic Grain Consumed Domestically by Humans and Animals

Region (poorest to wealthiest)	Share of Nutrients from Grainland Consumed Domestically by Humans	Share of Nutrients from Grainland Consumed Domestically by Animals
Africa	83	15
Asia	80	16
Latin America	49	39
European Union	28	41
North America	22	48

Source: See endnote 23.

countries would be exceedingly complex. But if one focuses on grain in selected countries and regions, the essential features of nutrient movements become clearer. Grain provides more than half of the calories ingested directly by most humans, and data on grain use and trade is reliable, making grain a revealing and manageable proxy for nutrient flows in general.[22]

In most countries, grain nutrients flow predominantly from farms to the nation's own people, rather than to animals, industry, or other countries. In developing countries, for example, more than three quarters of the grain produced is consumed domestically as food. (See Table 1.) And this share rises in less complex economies. Sub-Saharan African nations, for example, use 97 percent of the grain they grow for food. (By contrast, direct human consumption of grain in the United States accounts for only 28 percent of the total grain flows, the smallest of all U.S. grain nutrient trails.) These poorest nations produce virtually no exportable surplus, and their animals are largely pasture- rather than grain-fed, leaving nearly all of the grain harvest for domestic human consumption. This simple, rural-to-urban flow of nutrients would require an equally simple return flow to close the nutrient loop. Indeed, in most developing countries the recycling challenge is to return human and municipal wastes from cities to agricultural lands, a task made more difficult by the widespread absence of sanitation systems.[23]

With greater prosperity, people tend to eat more meat, and nutrient flows become more complex as grain is diverted to animal consumption. Impoverished Sub-Saharan Africa, for example, feeds only 2 percent of its grain to animals, but in the United States, 41 percent of grain nutrients go to animal consumption. (See Table 1.) Indeed, more U.S. grain nutrients are fed to animals than are consumed by Americans, by people in other countries, or by industry. Thus, in wealthy nations, nutrient recycling involves not only human and industrial wastes, but large volumes of animal wastes as well.[24]

Finally, some countries export a considerable share of

TABLE 2

Share of Nutrients from Domestic Grain That Is Exported: Seven Largest Grain Exporters

Country	Share of Nutrients Leaving Grainland for Other Countries
Australia	67
Argentina	46
Canada	45
France	44
United States	37
Thailand	31
Vietnam	12

Source: See endnote 26.

their nutrient outflow, which complicates recycling possibilities still further. These are countries with large productive capacity relative to domestic demand, and their ranks include both wealthy and developing nations. Among the world's top seven grain exporters, which includes nations as diverse as Canada and Vietnam, exports of grain nutrients range from 12 percent to 66 percent of domestic production. Unlike flows to the domestic populace and to animals, exported nutrients are largely unrecoverable by the exporting nation, although reuse in the recipient country is possible.[25]

In all, 10 percent of the world's grain nutrients flow across borders in grain; the figure would be somewhat higher if the grain content of exported meat were included in the analysis. (See Table 2.) As economies become increasingly integrated, and if import dependence grows, the volume of crop nutrients crossing national borders will rise. For net food exporters, the nutrient deficit is covered by using fertilizer. But even net food importers—who are accumulating nutrients from natural sources—often resort to heavier than necessary fertilizer use because they do not recycle organic wastes, or because getting organics back to farms is too expensive or difficult.[26]

TABLE 3
Net Flows of Nutrients in 15 Foods, by Region, Mid-1980s

Region	Magnitude of Net Flows (kilotons of nutrients)
Net Nutrient Importers	*Net Inflows*
Africa	533
Europe	2,809
Asia	1,034
Former Soviet Union	981
Net Nutrient Exporters	*Net Outflows*
North & Central America	3,387
South America	1,566
Oceania	225

Source: See endnote 27.

Research from the mid-1980s that focused on a larger set of commodities gives an idea of the net regional flows of nutrients. Tracking nutrients in 15 sets of foods, including grains, researcher G.W. Cooke found a large net shift out of the Americas and Oceania and toward the rest of the world: Africa, Europe, Asia, and the former Soviet Union. (See Table 3.) Perhaps more remarkable was the relative imbalance (inflows compared to outflows) for each region. The smallest relative imbalance was found in Asia, which nevertheless imported four times as many nutrients in food as it exported. North and Central America, by contrast, exported 76 tons of nutrients for every one it imported. Cooke's data demonstrate that nutrients in food flow across regions in highly skewed quantities.[27]

A heavy flow of nutrients in food into a nation does not mean that its farm soils are well supplied, however. Africa is a case in point. The continent takes in six times more nutrients in food than it sends out, but the soils of many African farms are steadily losing nutrients, thus exacerbating their need for imported fertilizer. Nutrients in food imports (like nutrients in domestic supplies of food) do not make their way to farm soils, but wind up instead in land-fills or at the bottom of rivers or bays. Thus, even net nutri-ent importers turn to fertilizer to replenish their soils, or—as in many African countries—watch soil fertility slowly decline.[28]

The use of manufactured fertilizer is the standard way to raise soil fertility in much of the world, and it is what allows large imbalances in food nutrient flows to be ignored. But fertilizer is often applied more liberally than necessary for plant growth (Sub-Saharan Africa is a notable exception), usually to ensure that crops are not underfed. Indeed, in the United States between 1991 and 1995, close to 56 percent more fertilizer was applied to grainland soils than left those soils in crops. (See Table 4.) In China, overapplication appears to be even higher, with nearly three quarters of the fertilizer applied unaccounted for in harvested grain. Although some of the excess is building up in farmland in the short run, a large share is leached or eroded away, and is responsible for the water pollution and ecosystem degrada-tion associated with heavy fertilizer use.[29]

Overuse of manufactured fertilizer could be reduced and soil quality raised if nutrient outflows were reused on farmland. The "waste" flows from food, feed, or exports are all potentially circular. Food becomes human excreta or garbage, which can be returned as sewage or composted food wastes. Feed becomes animal waste, which is applied to soils as solids, as liquid slurry, or as compost. Exported food and feed can follow similar paths once they reach their des-tination country. And these reused nutrients can be aug-mented using other wastes that did not originate on the farm, such as leaves and grass clippings. As it is now, how-ever, most of these branch paths are only partially looped back toward agricultural soils, if the loop is established at all.

The most widely recycled nutrients from crops are those in animal manure. For millennia, manure from cattle, pigs, poultry, sheep, and other farm animals has served both as a convenient and plentiful source of nutrients for plants, and as a tool for improving soil structure. Manure is still widely recycled to agricultural soils close to where it was produced. But in some countries, environmentally safe recycling is a growing challenge. As livestock operations become more centralized, manure is measured in hundreds of tons or thousands of cubic meters, and farmlands near these operations are challenged to absorb all of the waste that is produced.[30]

Recycling of human waste varies widely by region. Many Asian nations have long re-incorporated human wastes into farm soils, but the practice is on the decline. On the other hand, recycling of sludge and wastewater is on the rise in many sewered countries. The United States and Europe recycle a quarter to a third of the sludge they produce. While increasingly common, application of sludge to agricultural land is also controversial; sewage typically includes industrial as well as household wastes, and often contains heavy metals, toxic organic matter, and pathogens that are dangerous to human or environmental health. Thus the growth in recycling of human wastes is not always a positive trend.[31]

In many countries, municipal solid waste is a readily available but largely untapped source of nutrients and organic matter that could enrich soils. Organic material accounts for more than a third of urban wastes in industrialized countries and well over half in many developing countries. Only a small portion of this material is returned to soils: OECD member states composted just over one tenth of their organic wastes in the early 1990s. In developing countries, the potential for recycling is also largely unrealized.[32]

Greater reuse of organic matter on farms will not eliminate the need for outside sources of nutrients. Extensive nutrient losses are inevitable. The share of nitrogen in manure or sewage that is returned to the atmosphere through volatilization, for example, can be large—even as much as 50 percent (although these losses can be minimized through careful management of wastes). In addition, the high-yielding crop varieties in use today require more fertil-

TABLE 4

Efficiency of Fertilizer Use on Grainland: United States, China, and World

	United States	China	World
Share of applied fertilizer in harvested grain	64	27	53
Share not taken up by grain (calculated)	36	73	47

Source: See endnote 29.

izer than native varieties did. Still, reuse of organic matter can reduce the need for manufactured fertilizer while building soil fertility and health. In the process it can also help solve a surprisingly wide array of problems, from leaching and erosion to waste disposal.[33]

Composting Urban Wastes

The world's cities generate tons of natural wealth daily in the organic garbage—food scraps, yard trimmings, and paper wastes—that every household and many businesses and institutions throw away. This garbage is rich in organic matter—an essential ingredient for healthy soils—and it contains a modest supply of plant nutrients. Instead of exploiting this resource, however, most cities are intent on burying or burning it, or dumping it into rivers, lakes, or the sea. But as the benefits of reusing such material become evident, more cities are reclaiming it. To do so, they are turning to an ancient practice—composting—as a natural way to prepare the "waste" for reuse.

All organic materials contain both organic matter and nutrients. But working raw organic materials directly into the soil is not always the best way to exploit its organic matter and release its nutrients. Nutrients in materials that decompose slowly, for example, are "locked up" and unavailable for plant use. And in some soils, decaying organic matter can tie up soil nitrogen that would otherwise fuel plant growth. Fortunately, organic material can be converted—through composting—to a stabilized product that builds soils and releases nutrients in a steady and environmentally healthy way. Composting is a several-month-long process in which bacteria, worms, and other organisms feast on piles of carbon-rich matter and digest it, leaving behind humus, a rich, stable medium in which roots thrive. Worked into farm soils, humus builds soil structure and provides a productive environment for plants and essential soil organisms.

The ingredients for compost can come from a variety of sources. Food scraps, yard trimmings, paper, and sewage are all compostable, but most of this material is currently discarded. Food scraps and yard trimmings alone account for more than a third of the municipal waste flow in industrialized countries and well over half in many developing countries, which can afford fewer throwaway items. (Low-income countries have relatively small waste flows, but a large share of these flows is organic waste.) Yet, like OECD member states, most countries return only a small portion of this material to soils. (See Table 5.)[34]

If paper is included in the analysis, the compostable share of municipal solid waste jumps to more than 50 percent in industrial countries. Paper is best recycled into paper, not compost, but under certain conditions it is appropriate for composting. Where organic material is deficient in carbon, for example, paper can be added to raise its level. And when the market for recycled paper is saturated, composting paper can help to maintain the value of recycled paper. Had surplus paper been composted in 1996, when recycling centers were inundated, it would have stabilized paper prices and eased pressure on landfills and incinerators—in addition to returning organic matter and nutrients to farm soils.[35]

Beyond its contribution to waste reduction, the long-

TABLE 5

Composted Share of Organic Wastes, Selected OECD Countries

Country	Composted Share of Organic Wastes (percent)
Portugal	39
Spain	25
Denmark	23
France	19
Netherlands	15
United States	13
Sweden	10
Austria	10
Luxembourg	6
Belgium	5
Finland	5
Norway	4
Canada	3
Hungary	2
Poland	2
AVERAGE	11

Source: See endnote 34.

run value of compost lies in its capacity to build soils. Because it is riddled with pores, the humus in compost shelters nutrients and provides extensive surface area to which nutrients can bond; indeed, humus traps three to five times more nutrients, water, and air than other soil matter does. These characteristics also help retain nutrients that could otherwise be leached or eroded away. Thus, adding organic matter to soils further reduces the need for additional nutrient applications.[36]

Another important contribution of compost—suppression of plant diseases—has only recently been extensively documented. Since the 1970s, field tests have shown that compost limits the spread of root rot as effectively as many fungicides. Indeed, horticulturalists have found that compost in potting mixes makes fungicidal drenches largely unnecessary. Harry Hoitinck, a plant pathologist at Ohio State University and a pioneer in disease suppression research, asserts that compost use by nurseries in Ohio has eliminated the use of methyl bromide—a potent fungicide highly poisonous to humans, and an ozone-depleting substance whose use is soon to be banned. Because chemical alternatives to methyl bromide are less effective or are also unsafe, the disease-suppression capacity of compost is welcome news. Scientists are now learning to augment this capacity by inoculating compost with beneficial organisms.[37]

Compared to its advantages for soil building, water retention, and disease suppression, the nutrient contribution of composted urban organic material is modest, but significant nonetheless. Nutrients in municipal solid waste (not including paper) in OECD countries amounted to an average 7 percent of their commercial fertilizer use in the early 1990s. (See Table 6.) Because fertilizer is commonly overapplied, however, the potential contribution of urban nutrients is actually larger than the 7 percent figure indicates. If fertilizer use in OECD countries were reduced by a third—less than the rate of nutrient overapplication in many industrialized countries—nutrients in solid waste would amount to 12 percent of nutrients applied as fertilizer. This level of nutrients (which does not yet include those available from human waste) begins to offer potential for cutting the pollution of water caused by fertilizer overuse.[38]

How much reduction in fertilizer use is allowed by incorporation of compost depends on the makeup of the compost, the amount applied, soil and climate conditions, and the crops being cultivated. Compost can reduce fertilizer use because of its own nutrient contributions, but also because of its capacity to reduce leaching, which allows a greater share of applied fertilizer to be used by plants. On the other hand, the fact that compost releases its nutrient supply very gradually (unlike fertilizer, whose nutrients are immediately available to plants), only allows the full nutrient contribution of compost to be realized over time, after soils have been built. Still, compost use has already led to

TABLE 6

Nutrients in Organic Municipal Solid Waste (paper excluded) as a Share of Fertilizer Use, Selected OECD Countries

Country	Nutrients in Organic MSW as Share of Commercial NPK Use
Mexico	17
Turkey	15
Japan	14
Netherlands	12
Belgium-Luxembourg	11
Italy	9
Portugal	9
Switzerland	9
Australia	8
Spain	7
Austria	6
Canada	6
Sweden	6
Finland	5
Greece	5
United States	5
Norway	4
Denmark	3
France	3
AVERAGE	8

Source: See endnote 38.

reductions in fertilizer applications in some areas. According to a World Bank report, for example, farmers in India who use a commercial compost called Celrich cut chemical fertilizer consumption by some 25 percent.[39]

Finally, composting is accessible to people who are poor. Because it is a decentralized and natural source of wealth—every household produces composting ingredients—it can promote better nutrition among the urban poor who cultivate their own food. An estimated 200 million city dwellers worldwide now practice urban agriculture, supplying part of the food needs of some 800 million people. In

Kampala, Uganda, for example, 35 percent of households produce their own food. And in Accra, Ghana, urban residents supply the city with 90 percent of the vegetables consumed there. For the urban poor, compost is a virtually free fertilizer and soil builder, whose production requires little space, virtually no equipment, and a modest amount of labor. Such a valuable and affordable resource, available without reliance on outside suppliers, can make a large economic and nutritional difference to people living on the economic margins.[40]

For all its wonders, compost presents some important managerial challenges. Composts vary from place to place—and even from batch to batch—because the combination of inputs can vary so widely. Yard clippings are more available in summer than in winter, for example, and their nutrient make-up changes with the seasons. Paper availability may depend on the ups and downs of the economy. The good news is that this complexity allows composts to be tailored to the particular soils and crops they will serve. But it also requires that compost makers know their customers and respond to their diverse needs, and that users understand how the product works in soils. Creating the right compost for a particular use and employing it optimally will require more research and outreach than is typically available today.[41]

As the many advantages of composting become apparent, its practice is taking off. In the United States, composting facilities multiplied more than fourfold between 1989 and 1996, from some 700 to more than 3,200. Many cities and counties now make organic matter available to the public for use as mulch, or as the feedstock for compost making. In San Jose, California, a recently completed three-year pilot program to promote the use of compost led to a 54 percent increase in its production by local processors, and demand for the product was brisk.[42]

Compost is increasingly recognized as good business. Evidence of this is the experience of Community Recycling of Southern California, which saw gold in the spoiled fruits and vegetables of area supermarkets. The company mixes the produce with yard wastes from the area to generate 125,000 tons of compost per year—the maximum allowed under its permit. Today, the two largest supermarket chains in southern California, representing more than half of the grocery outlets in the region, have their organic wastes composted by this firm.[43]

Community Recycling is not the only beneficiary of its organic recycling program. The compost is spread over some 12,000 hectares of farmland, whose soils enjoy the multiple benefits of higher levels of organic matter. Farmers profit directly too: the company calculates that nutrients in one ton of its compost would cost $58 if purchased as fertilizer. But the company sells its compost for $10 per ton.[44]

Compost is a virtually free fertilizer and soil builder.

Composting also has economic benefits for institutions that generate organic waste. Some schools, prisons, hospitals, and other food-serving establishments save money by having food scraps composted instead of hauled away for disposal. Middlebury College in Vermont, for example, reports annual savings of some $25,000 by sending food residuals to a compost facility rather than to a waste disposal operation. The New York State Department of Corrections has saved more than $1 million by composting food scraps at 31 sites around the state. The key is for composters—whether individual farmers or large, centralized operations—to charge less to accept the organic material than a landfill or other disposal destination would. The generator of the waste saves on disposal costs, and the composter receives revenue to haul away material that will be transformed into a profitable product. The "win-win" possibilities of composting are indeed extensive.[45]

Such mutual advantages, however, are not automatic or guaranteed. The Indian government, for example, has tried several times in recent decades to promote composting of municipal wastes, but the schemes have largely failed, for various reasons. Inputs to the composting process were not

well monitored, and inclusion of non-organic material low-
ered the quality of the resulting compost. Poor equipment
maintenance led to breakdowns and inconsistent production. City governments were seldom committed to the federal government's vision of widespread composting. And
subsidies on fertilizer made compost economically uncom-
petitive. While the potential benefits of composting are
manifold, the Indian experience demonstrates that effort is
required to avoid a number of potential pitfalls.[46]

It is ironic that composting, so lately embraced in
many economies, is one of the oldest forms of recycling
known to humankind. As societies become reacquainted
with this practice, its value as a natural solution to problems
from overflowing landfills to anemic soils will become
apparent. Then, with the proper institutional and economic
incentives, composting could become as commonplace as
the recycling of cans, newspapers, or paper is today.

The Potential and Peril of Human Waste

Most of the world's cultivated food passes through
human beings, so it is no surprise that human waste is
a trove of nutrients and organic matter. Harvesting this
material for agriculture is a natural way to close an important organic loop; indeed, Chinese farming thrived on recycled excreta for thousands of years. But as more cities
process these wastes using technologies designed to dispose
of them, rather than reuse them, safe recycling of human
waste becomes much more difficult. Safe reuse is best
ensured by shifting away from disposal technologies—such
as conventional treatment plants, or sewers that mix industrial and domestic waste—and toward technologies engi-
neered to produce a clean fertilizer. For countries not yet
committed to expensive disposal systems, this shift can
occur more quickly than for those that are. Until such a shift
takes place, the reuse of human excreta can be safely prac-

ticed only by observing the strictest standards.[47]

Most excreta is not reused, although reuse—often unsafely practiced—is growing. In developing countries, where 72 percent of the population has access to adequate sanitation, sewers, septic systems, and pit latrines are the dominant disposal systems. Sewers and septic tanks predominate in Latin America and the Middle East, while Africans and Asians rely at least as heavily on pit latrines. Most sewers flow to the nearest river, bay, or ocean; only 10 percent of this sewage receives treatment. Where pit latrines are used, waste material typically remains buried. Except for parts of Asia, which has a long history of excreta reuse, and some arid regions, where sewage water (often untreated) is commonly used for irrigation, human waste is widely regarded as unwanted debris.[48]

Industrial countries have long had the same perspective, but this is changing. Many now encourage reuse of sewage sludge on farmland, and the practice is growing. European countries applied roughly one third of their sewage to agricultural land in the early 1990s, while the United States applied 28 percent. The growing interest in reuse may reflect dwindling options for cheap disposal, rather than a strong interest in building farm soils. Traditional dumping sites—landfills, incinerators, and oceans—are less available, more costly to use, or legally off-limits today, while farmland is often an inexpensive alternative disposal site. But just as sewers and treatment facilities are not designed for recycling, farmland is not suited to absorb the chemicals and heavy metals often contained in the sewage stream.[49]

If human wastes *are* made safe for use on farmland, however, their reuse can help reduce applications of chemical fertilizer. In many developing countries, the nutrient content of human waste is equal to a substantial share of the nutrients applied from fertilizer, even after losses of nitrogen to volatilization are taken into account. (See Table 7.) For OECD countries, nutrients in human waste that is not already spread on land equal roughly 8 percent of the nutrients applied as fertilizer. As with municipal organic waste, this figure understates the potential contribution of nutrients in human waste. If fertilizer use in OECD countries were reduced by a third, nutrients in human waste would amount to 12 percent of nutrients applied as fertilizer.[50]

Recycling human waste, however, will require different technologies, or different ways of using existing ones. Modern methods for *disposing* of human waste are not designed for *reusing* it. Sewers, for example, commonly serve residences and industry together, a practice that often contaminates organic matter with heavy metals or toxic chemicals. Conventional treatment plants are designed to remove nutrients (and other matter) from wastewater, which lowers the enrichment level of effluent used for irrigation. Moreover, conventional treatment methods (with the exception of disinfection, which is rarely practiced in developing countries) reduce pathogens by too little for safe reuse in agriculture. Thus, many of today's disposal technologies are not suited to produce fertilizing products.[51]

Where sewers and treatment plants have been turned to waste reuse, there have been mixed results, at best. Even in countries considered successful with reuse—Israel, for example, which diverts treated wastewater to irrigation— caution is warranted. The country began large-scale reuse of sewage effluent in 1972, and today recycles 65 percent of its wastewater to crops. No excessive rates of illness have been linked to its use. Nevertheless, cadmium levels have been shown to increase by 5 to 10 percent annually in Israeli effluent-fed soils, and heavy metals were found to have accumulated in an aquifer below land that was irrigated with effluent for 30 years. If industrial wastes were not dumped in sewers, the country could more safely apply sewage effluent to crops. Better yet, if human wastes were managed using dry (non-sewered) methods such as composting toilets, the water currently used to carry sewage would be available to agriculture as clean water.[52]

Where sewers are little more than feeder lines to irrigation canals, and where the sewage they carry is untreated,

TABLE 7

Nutrients in Human Waste as a Share of Nutrients in Fertilizer Applied, Selected Countries

Country	Nutrients in Human Waste as a Share of Nutrients in Fertilizer Applied (percent)
Kenya	136
Tunisia	52
Indonesia	49
Zimbabwe	38
Colombia	31
Mexico	31
South Africa	29
Egypt	28
India	26

Note: assumes loss of 50% of nitrogen content to volatilization.
Source: See endnote 50.

risks to human health are much greater. Raw sewage used to irrigate vegetables and salad crops is blamed for the spread of worm-related diseases in Berlin in 1949, typhoid fever in Santiago in the early 1980s, and cholera in Jerusalem in 1970 and in western South America in 1991. Even so, the risky use of wastewater continues in many developing countries. In the Mexican state of Hidalgo, wastewater from Mexico City is used in the world's largest wastewater irrigation scheme, covering some 80,000 hectares. The effluent, which is 55-80 percent raw sewage (the balance is storm water), is barred from use on some salad crops, but other foods, including corn, wheat, beans, and some vegetables, are irrigated with sewage water.[53]

In contrast to wastewater reuse, application of sludge to farmland carries a different set of risks, especially where industrial wastes or household chemicals are part of the sewage flow. Researchers from Cornell University and the American Society of Civil Engineers have found more than 60,000 toxic substances and chemical compounds in U.S. sewage sludge, and report that 700-1,000 new substances are developed every year, some of which also enter the sewage stream. These substances include PCBs, pesticides, dioxins, heavy metals, asbestos, petroleum products, and industrial solvents, many of which are linked to ailments ranging from cancer to reproductive abnormalities. They are also a threat to soils: once introduced to cropland, for example, heavy metals persist for decades (as in the case of cadmium) or even centuries (as in the case of lead). Because little control is exercised over what enters sewers, the contents of a given load of sewage sludge can be highly unpredictable and potentially dangerous to people and soils.[54]

Although industrialized nations maintain standards for sludge reuse, these may be lax. Such standards in the United States are the least stringent of any in the industrialized world, with allowable levels of heavy metals an average eight times higher than in Canada and most of Europe. Indeed, Cornell University researchers have recommended that U.S. farmers apply sludge at no more than one tenth the levels permitted by the U.S. Environmental Protection Agency. Moreover, testing in the United States is required infrequently—as seldom as once a year for the smallest applied amounts—even though the contents of sludge can vary greatly from load to load.[55]

Clearly, reliance on mixed-waste sewers and treatment plants, the "modern" way to process human waste, does not guarantee output that is safe for use in agriculture. Other technologies, most of which are simpler and cheaper than sewers and treatment plants, may offer greater possibilities for recycling wastes. Indeed, opportunities exist for developing countries to "leapfrog" past industrial nations by adopting cutting-edge technologies that are affordable and environmentally sound, and that help to close the organic loop by safely returning human wastes to agriculture.

One simple—and ancient—alternative to sewage treatment plants is waste stabilization ponds, a series of holding areas in which sewage is retained for 10 days to a few weeks. Bacteria and algae work to convert the effluent to a stable form as it passes from pond to pond. Stabilization ponds require more land than conventional treatment plants, but they are much cheaper, simpler to build and maintain, and, best of all from a recycling perspective, more effective at producing safe irrigation water. A conventional treatment plant can reduce the number of fecal coliforms in a milliliter of water from 100 million to 1 million, a 99 percent reduction—but not enough for use on crops. For unrestricted irrigation use, the World Health Organization recommends a fecal coliform level a thousand times lower—no greater than 1,000 per milliliter—and waste stabilization ponds can achieve this.[56]

One variant of the waste stabilization pond is a wetland modified to process wastes, the showcase example being the one in Calcutta. For more than half a century, sewage has been channeled to a wetland east of the city, where multiple ponds are used not only to process waste, but also to raise fish and provide nutrient-rich irrigation water for farmers. The system works by mimicking the interconnectedness of a natural ecosystem. Nutrients in the waste feed fish, plants, and organisms in the ponds. The fish, in turn, greatly reduce or eliminate algal blooms, making the final wastewater product more useful for agriculture. Water hyacinth cultivated at the ponds' edges further purifies the water and protects the banks from erosion. And the hyacinth is either harvested for animal feed or composted. These multiple benefits, combined with a cost less than a quarter that of a conventional sewage treatment plant, have made the area a valuable municipal resource.[57]

Opportunities exist for developing countries to "leapfrog" past industrial nations by adopting cutting-edge technologies.

A constructed micro-version of the Calcutta wetlands system could provide waste-processing capacity for some industries, thereby preventing their wastes from entering the sewer system. Complete with plants, microorganisms,

and even fish, these facilities consist of a series of pools and
constructed wetlands, often built in a garden-like setting,
which progressively treat industrial wastes. One U.S. firm
has found a robust market for these facilities, with 20 projects built or under construction since 1992 at businesses
and institutions as diverse as the M&M/Mars Company in
Brazil and Oberlin College in Ohio.[58]

For all their advantages, these natural filtering systems
are land intensive. Stabilization ponds are estimated to
require 30 hectares for every 100,000 people served. And
Calcutta's wetlands system required 3,200 hectares to
process roughly a third of the city's wastewater in 1991. The
industry-level facilities also require an extensive area, which
may prove prohibitive in crowded cities. Where land is
tight, other choices are available, some of which can avoid
the expense of sewage infrastructure.[59]

One of the more promising options for processing
sewage safely is a series of simple technologies developed
and patented in Mexico and known collectively by their
Spanish acronym, SIRDO. SIRDO systems build on the "dou-
ble-vault" waste treatment concept developed in Vietnam,
under which one chamber collects current deposits of waste
while the other is closed for several months as previously
deposited material composts. Solar heating and bacteria
transform wastes and other carbon matter into a safe and
odorless "biofertilizer" that is sold to nearby farms.[60]

SIRDO technology is applied in diverse ways. Some
designs are "dry," requiring no water—and no sewage infrastructure—for their operation. Dry units are self-contained
structures that are detached from a house and serve one or
two families. They compost household organic matter
together with human waste, thereby easing pressure on
landfills and sewage treatment plants. "Wet" SIRDO units
are neighborhood-level mini-plants that biologically process
the wastes of up to 1,000 people, operating in conjunction
with existing flush toilets and local sewer lines. Even these
"wet" systems are water savers, because they separate grey-
water from solids and percolate it through a bed of sand and

gravel until it is purified enough to reuse on gardens, or to irrigate non-food crops. These systems are simple enough to maintain and operate that they do not require constant oversight by an engineer. A trained lay person can handle day-to-day operations, with occasional assistance from a SIRDO specialist. Several of the wet units in Mexico City are maintained by the gardeners of the condominium complexes in which the units are located.[61]

SIRDO's advantages extend beyond fertilizer production and water savings. As an effective sanitation technology, SIRDO improves the level of public health by reducing illnesses caused by pathogen-tainted water supplies. In the warm climates where SIRDOs are currently used, the unit's solar-heated waste chambers generate higher temperatures, over longer periods, than are needed to ensure that pathogens are killed. In the town of Tres Marias, Mexico, introduction of SIRDO technology and a new potable water system are credited with cutting the rate of gastrointestinal illness from 25 cases per person in 1986 to less than one case per person in 1990. Since contaminated water is a major cause of sickness and death among children in developing countries, the technology's success in sterilizing wastes is a welcome advance.[62]

Moreover, the SIRDO systems are affordable, and they even generate modest flows of revenue. A cost-benefit analysis undertaken by the National Wildlife Federation (NWF) found that all five SIRDO models studied—three wet and two dry—offered net financial gains under Mexican market conditions for water, labor, and bio-fertilizer. The simplest dry design, for example, costs $307 for set-up and $20 per year for maintenance—a total of $607 over 15 years—but earns the owner $2,088 in fertilizer revenues in the same period. The net income for user families is modest—on the order of $30-60 dollars per year—but nonetheless meaningful for people living on the economic margin.[63]

Significantly, the NWF analysis was limited to private costs and benefits. It did not consider the technology's social benefits, which include the reduced need for sewage treatment, boosted levels of public health, and improved soil structure and fertility on farms that use the bio-fertilizer. SIRDO's multiple advantages to users and society have spurred its adoption in Guatemala, Chile, and nine states in Mexico.[64]

Another non-sewer approach to waste processing doubles as a source of energy. Since the 1970s, China has installed more than 5 million anaerobic digesters—large chambers, sitting mostly underground, that break down a rural family's organic waste, including manure, human excreta, and crop residues, producing gas in the process. Toilets and pigsties drain directly into the digester, which yields enough biogas to meet 60 percent of a family's energy needs, mostly for cooking and for fueling gas lamps. The unit also produces an odorless dark slurry, used primarily for fertilizer, but also viable as feed for livestock or fish. The digesters are inexpensive—$80 covers the cost of materials and the help of a technician in construction.[65]

In cities that are already sewered, and whose populations are accustomed to flush toilets, separation of human and industrial wastes will be more challenging, and may need to be viewed as a medium- to long-term goal. Nevertheless, current technologies suggest several possible approaches. Dry composting toilets, for example, can be installed in the bathrooms of many suburban homes. They look like standard flush models—without the water tank—and can hold up to several years' worth of excreta. They require some maintenance, including occasional additions of carbon material, such as sawdust or leaves, and periodic inspection of the equipment and the compost itself. Service contracts, however, can minimize the burden on homeowners. Other non-sewer technologies include micro-flush toilets, which use as little as one pint of water per flush, and vacuum-powered toilets similar to those in aircraft lavatories. All of these systems create a fertilizing product that can be applied to home gardens or, where economically feasible, collected and sold to farmers. And because the excreta is segregated from the flow of detergents, cleaning products, solvents, and other chemicals used in many households, the composted material is clean. The systems are not cheap, however, ranging in price from $1,000 to $6,000 per unit.[66]

Large buildings, such as multi-story apartment complexes, would be served with different technologies. (Composting toilets usually require that the holding chamber be located directly below the toilet, which makes their use in multi-story buildings impractical.) Constructed wetlands are one possibility for buildings that have plenty of land. A more viable option is the use of biogas digesters, similar in concept to those used by some Chinese peasants, but built on a larger scale. Located in the building's basement, the digester would collect wastes from standard low-flush toilets and produce two products: methane, which could provide part of the building's power, and uncontaminated sludge, which could be collected and applied to farmland. Digesters offer a glimpse of the multiple benefits possible from full exploitation of human "waste."[67]

The nutrients in human waste constitute a vast, untapped agricultural resource.

The nutrients in human waste constitute a vast, untapped agricultural resource. Getting them safely back to farmland would help to build soils and reduce the need for additional nutrients from fertilizer. But separating human excreta from industrial wastes—the prerequisite for safe recycling—will require imagination and commitment. Ironically, unsewered cities may be in the best position to capitalize on new technologies for excreta management, technologies designed to produce an uncontaminated fertilizer product.

Sustainability and Scale

Fertilizer and cheap transportation were the original scis-
sors that snipped open organic loops, thereby unleashing
the pollution and waste problems described earlier. Today,
the globalization and concentration of agriculture com-
pound these problems by stretching and fattening nutrient
flows. The surge in agricultural trade, for example, redistrib-
utes nutrients unevenly around the world, driving some
regions to a heavier-than-necessary dependence on fertilizer,
and leaving others with unhealthy nutrient surpluses. And
concentration of production can swell nutrient streams
until nutrient accumulations become unmanageable. The
emerging lesson is that scale matters, and that too large a
scale can lead to distortion and mishandling of nutrient
flows. Even the scale of recycling operations can determine
whether cities are successful in actually closing nutrient
loops. Prospects for restoring circularity to organic flows
may depend on limiting the scale of agricultural operations
and some recycling operations so as to shorten and unplug
today's linear movements.

The globalization of food flows may be the sleeper agricultural story of recent decades. The last 40 years are widely
heralded for their unprecedented growth in output, but agricultural trade—and the displacement of soil nutrients that
trade entails—grew even faster than production. World
grain output, for example, doubled between 1960 and 1995,
but grain exports *tripled* during the same period. Indeed,
growth in agricultural trade has outpaced production con-
sistently since 1960, except for a short period in the mid-
1980s. Today, more dinner plates are filled with food of dis-
tant origin, and more nutrients cross national borders, than
ever before.[68]

The uneven redistribution of food nutrients resulting
from increased international trade generates net losses in
some areas, and net gains in others. Several countries in
northern Europe, for example, suffer from excessive accu-

mulations of nutrients, many of which are imported across oceans. An extensive European livestock industry purchases feed from as far away as Brazil, Thailand, and the United States. But the industry has outgrown the capacity of nearby lands to absorb its wastes, so manure has steadily accumulated. Indeed, early this decade, the Netherlands could boast the world's largest "manure mountain"—some 40 million tons' worth.[69]

These accumulations, coupled with heavy fertilizer use, are responsible for serious pollution problems in the Netherlands. Nitrate levels in the country's groundwater were more than double the recommended maximum level in the early 1990s. So saturated was the country in nitrogen and phosphorus at mid-decade that farmers could have met their crops' nutrient requirements from manure alone— without a single application of nitrogen or phosphorus fertilizer—and still ended up with a nutrient surplus in their soils. The mismatch between the scale of activity—heavy flows of nutrients from three continents that converge on a single small region—and the environment's limited capacity to absorb the output of that activity demonstrates the relevance of scale. The inflow of nutrients to the region was so large that they could not be recycled there, nor could they be returned to their original soils.[70]

Taiwan finds itself with similar problems, after building a substantial, but import-dependent, hog-raising industry. The country buys more than 90 percent of its corn feed from farmers in the midwestern United States, an ocean and half a continent away. But the oversized hog-raising industry produces more manure than the country can handle, resulting in extensive pollution, as in the Netherlands. Indeed, officials in Taiwan estimate that two thirds of Taiwan's water pollution is the result of manure discharges from hog farms. As a result, the government has been struggling since 1991 to reduce the number of hogs by one third.[71]

Lengthened nutrient supply lines are also found within countries. This is especially clear in the United States, where feed is shipped ever greater distances as cattle-, hog-,

and chicken-raising facilities move away from feed production regions. Cattle feedlots, for example, were once located
in the Corn Belt states that supplied them with feed, but
they began to move hundreds of kilometers west to the
Great Plains in the 1950s and 1960s. More recently, hog production has shifted hundreds of kilometers east, from the
Midwest to Virginia and North and South Carolina. The supply line connecting feed and livestock, which once extended a few hundred meters from field to barnyard on a single
farm, has now been stretched across state lines, essentially
precluding the return of nutrients to feedcrop fields.[72]

The separation of livestock from crops is linked to
another issue of scale, the size of agricultural operations.
Large operations realize economies of scale that allow them
to absorb the transportation costs resulting from long-distance food and feed shipments. Not surprisingly, then, the
shift of livestock production away from feed-producing
regions in the United States was accompanied by an increasing concentration of operations. (See Table 8.)[73]

As livestock operations have centralized, however, so
has manure, creating a waste disposal dilemma where farmers once saw only a resource. Indeed, facilities with tens of
thousands of animals measure their waste production in
hundreds of tons of manure or thousands of cubic meters of
slurry. The slurry lagoon on one mega-farm in Missouri, for
example, covers 2.8 hectares, is 5 meters deep, and holds
more than 87,000 cubic meters of effluent. Such large accumulations of waste cause serious environmental damage: the
U.S. Environmental Protection Agency reports that effluent
from centralized livestock facilities accounts for more than a
quarter of the water pollution caused by agriculture in the
United States. In North Carolina alone, over half a dozen
major lagoon spills were reported in 1995, including one
involving some 95,000 cubic meters of lagoon effluent.[74]

The buildup of nutrients from large animal-raising
facilities has been documented in several states. In
Delaware, for example, farmers were applying 72 percent
more nitrogen and phosphorus than their crops needed in

TABLE 8

Concentration of Livestock Production in the United States

Livestock	Degree of Concentration
Beef	More than a third of marketed cattle come from just 70 of the nation's 45,000 feedlots. The number of feedlots in the top beef-producing states has fallen by 75 percent over the past two decades. The largest facilities are found in Kansas, Nebraska, and Texas.
Poultry	97 percent of U.S. sales are now controlled by operations that each produce more than 100,000 broilers per year.
Pork	U.S. hog and pig inventory has climbed some 18 percent in the past decade, while the number of operations has decreased by 72 percent. In North and South Carolina and Virginia, where the industry is increasingly located, nearly 80 percent of hogs come from facilities with 5,000 or more head. In traditional hog-raising states, only 6 percent of hogs come from facilities of this size.
Dairy	The number of dairies has fallen from 250,000 a decade ago to 150,000 today, and average herd size has increased by more than 50 percent.

Source: See endnote 73.

the early 1990s, thanks in large part to heavy use of manure from the state's extensive poultry-producing operations. Without enough cropland to absorb the mountains of manure generated each year, the material is applied wherever possible, at a rate of overapplication that averages some 50 kilos per hectare. The manure generated by poultry could meet well over half of the state's crop nutrient needs if it could be easily and economically distributed, but large, centralized operations make this difficult.[75]

The case of burgeoning livestock facilities demonstrates that operations can be too large relative to the absorptive capacity of the surrounding environment. In other cases, operations are too large relative to a region's base of technology or skilled labor. In many developing countries, for example, composting municipal solid waste is more likely to occur—and recycling is more likely to be realized—if it is pursued in a decentralized way. Centralized facilities often require expensive imported machinery, which eliminates jobs and requires technical skills that may not be available. Neighborhood-level composting, by contrast, requires less capital investment, creates more jobs, and can use simple methods that rely on nature to do the composting work.[76]

Consider the history of composting in India. Despite a commitment by national governments as early as 1944 to increase composting and recycling in major Indian cities, most initiatives have been spectacular failures. Projects were designed to handle large quantities of organic material—between 150 and 300 tons per day—and were supplied with imported machinery for large-scale processing. But the machinery was poorly maintained and workers were inadequately trained. Moreover, the garbage fed into it contained large amounts of non-organic material—70 percent of the compost ingredients in Calcutta, for example—which damaged machinery. These problems raised operational costs, and with little effort to market the final product, cities soon found themselves operating financial sinkholes.[77]

Small-scale, low-tech composting efforts in other developing countries offer an encouraging contrast to the Indian experience. The city of Coimbra, Brazil, for example, wanted to reduce inflow to its dump, which was leaching contaminants into a stream used for irrigation and as a water supply for animals. Using local materials and local labor, the city built a simple facility that could process 15 tons of waste per day. Twelve handcarts and three animal carts collect refuse daily from the city's 7,000 residents. The organic material is then manually sorted and composted naturally in various piles at different stages of maturation. The center produces more than 12 tons of compost per month from waste that would previously have been buried. More important, the labor-intensive composting process provides work for eight people—five of whom had lived and worked as scavengers at the landfill.[78]

A still more decentralized process is used in Cairo by the Zabbaleen, a group of poor Coptic Christians who survive by collecting garbage from wealthy neighborhoods and using it as feed for pigs, which are raised in enclosed courtyards throughout the city. The Zabbaleen deliver the manure and other organics to a local composting facility. Because the manure is already partially decomposed, composting time is greatly reduced, to a week or two. The resulting compost is coveted by farmers within a 100-150-kilometer radius of Cairo, who pay for its delivery. The product is not perfect—it often has high levels of lead and zinc—but an effort to have households separate their organic garbage from other wastes before it is collected could largely eliminate this problem.[79]

The question of scale is often treated solely as an economic issue. From this limited perspective, bigger is better, because economies of scale typically make large operations more competitive than small ones. But equally relevant are "ecologies of scale," under which larger operations may be more environmentally damaging because they reduce the possibilities for successful recycling. These ecologies of scale provide a more complete picture of the costs and benefits of large, centralized operations.

Returning to Our Organic Roots

As the drawbacks of today's linear organic flows become evident, interest in recycling organic material is growing. But a return to time-honored recycling practices is not simple in an increasingly urbanized and industrialized world. Many economies are deeply invested in linear flows of organic matter, and will require time to reestablish organic loops. They will also have to wrestle with fundamental issues of sustainability, including the maximum sizes of viable cities and agricultural operations, and the maximum extent to which food should be traded and food raising should be concentrated. But commitment to a series of principles of organic matter management is a good first step; from these principles, specific policies can emerge to close the loop.

The baseline precept for organic matter management is this: *in a fully sustainable world, all organic flows must cycle.* By this principle, any instance of organic dumping—whether of garbage sent to a landfill or incinerator, sewage flowing to a bay, or manure overapplied to farmland—represents unacceptable waste of a natural resource. Just as policymakers and citizens would not tolerate the wanton burning, dumping, or burial of natural resources, neither would they allow organic matter to be casually discarded if they saw its true value. Appreciation of the contribution of organic matter to sustainable urban living will require a diverse set of policies affecting the individuals, municipalities, and industries that produce organic waste and the farmers that use it.

As a starting point, organic material can be turned away from traditional disposal sites using taxes or legal restrictions. The U.K., for example, has instituted a landfill tax designed to discourage landfill use, while several U.S. states have mandated cuts in organic inflows to landfills, or bans on particular kinds of organic matter, such as grass clippings. The U.S. has also banned ocean dumping of sewage, and Europe is set to do so as well. Each of these diverse policies closes another door on organic dumping.[80]

Outlawing dumping, however, is only half the battle. Viable recycling options are necessary to ensure that material is actually reused. Such options are best governed by two more principles, first, that *organic wastes should be segregated from other wastes.* It is generally simpler and cheaper to prevent contamination of organic material than to try to clean up dirty material. Segregation of wastes from the beginning is the best way to do this. Once this precept is accepted, a third principle can expedite the search for viable recycling options: *those who generate waste must recycle it, or pay for its recycling.* This variation of the "polluter pays" principle applies to individuals, businesses, and institutions alike, and spurs each to find the most efficient way to reuse material, and possibly to reduce its flow. City government can still play a large role in helping citizens and businesses to recycle, however.

Armed with these precepts, the search for viable recycling options can proceed on different levels. Municipal educational programs, for example, can equip residents to take responsibility for their food and garden wastes. Sonoma County in California has reduced landfill inflows through a citizen training program for composting, and participants have cut their landfilled wastes by an average of 18 percent. Best of all, this and similar programs are cost effective: they spend $12 for every ton of waste diverted, which amounts to less than 40 percent of what a landfill would charge to take the material.[81]

Where capital investments by individuals or institutions are required, an education program describing the potential financial gains could help to grease the wheels. Most institutions that generate large amounts of food wastes, such as hospitals, schools, or prisons, have likely given little thought to composting. Yet the experiences of Middlebury College and the New York State Department of Corrections cited earlier demonstrate that composting can make financial as well as environmental sense. Getting the word out could help jumpstart recycling by such institutions.

For human wastes, moving toward viable recycling technologies may be a long-term process, especially where cities are committed to *disposal* technologies. Again, education is the first step. Cities and citizens will have to determine whether their current system is capable of producing a clean fertilizing product. If not, exploration of alternative systems designed for recycling is warranted, along with plans for adoption of an appropriate technology. In developing countries—many of which need to invest in construction or maintenance of sanitation systems anyway—such a reassessment of sanitation represents an opportunity to leapfrog over the costly problems encountered by industrial countries with their "modern" systems. Funding limitations are cited as the chief obstacle to construction of sanitation infrastructure in low-income countries. Yet a system of composting toilets—an option that most cities do not consider—costs less than one seventh that of sewage systems. In any case, weaning a city of its sewers is not a quixotic notion. The entire province of Tanum in Sweden is converting to composting toilets, and is already enjoying the environmental benefits of the shift: nitrogen and phosphorus pollution has been reduced by 90-95 percent compared to the levels experienced when the region was sewered.[82]

Converting organic wastes to a useful product, however, does not ensure that organic matter will be recycled; farmers have to *want* to use it. One reason farmers are slow to choose organic materials over fertilizer is that they are uneasy about it. Synthetic fertilizer comes in specific formulations, with the amount of N, P, and K marked on the package label. But compost, sludge, and manure are often highly variable products made from equally variable inputs. Farmers may be unsure of how much to apply and at what rate nutrients will become available to plants. Indeed, markets for organic matter will not mature until farmers can be confident about the product they are buying, and until suppliers can respond to the diverse needs of different soils and different crops.

This level of sophistication will require greater research into the nature and properties of organic matter, especially compost, and how these function in different soils and climates, and with different crops. But research in organic agriculture receives little official support. An innovative investigative project by the California-based Organic Farming Research Foundation determined in 1997 that just 34 of the 30,000 research projects—one tenth of one percent—funded by the United States Department of Agriculture between 1991 and 1996 focused on organic agriculture. Moreover, no provision was made for dissemination to farmers of the results of these few projects. Such institutional indifference will need to be reversed if organic recycling is to make the maximum possible contribution to agriculture.[83]

Even if organic dumping were proscribed and farmers were eager to apply organic matter, sustainable organic flows would not necessarily be achieved. Nutrients might still accumulate in one area and be unnecessarily depleted in another. Here, a fourth principle emerges: *budgets for nutrients should be established to keep nutrient flows in rough balance*. Nutrient budgets are most meaningful at the farm level, and can be as useful to farmers as they are helpful to the environment. A simple tool known as the Nutrient Management Yardstick has been developed by the Center for Agriculture and the Environment in the Netherlands to track farm-level nutrient flows. The tool is a workbook that helps farmers to keep track of all nutrients brought onto the farm—whether in fertilizer, feed, manure, or other materials—and all nutrients that leave the farm in crops, livestock products, or other materials. Dutch farmers using the yardstick have registered reductions in nutrient surpluses in each of its six years of use, and its adoption is likely to be widespread as farmers develop mandated nutrient management plans starting in 1998. Dissemination of this simple tool through agricultural extension programs could be an inexpensive way to get a handle on nutrient flows at the source.[84]

Because nutrient flows are measurable, agricultural operations can be held accountable for safely maintaining nutrient balances. Indeed, any operation likely to have large on-site nutrient imbalances—like the massive nutrient inflows common to centralized livestock facilities—should have a plan for disposing of nutrient surpluses in a way that is environmentally healthy. Until a facility, for example, can demonstrate that nearby landowners are willing to receive its excess manure, and that the manure will be applied at rates that can be safely absorbed by those soils, the facility should not be allowed to expand.

Farmer understanding of nutrient dynamics is necessary to fulfill the fifth principle of organic cycling: *chemical fertilizer should supplement inflows of organic matter, and levels of application should not exceed the crop's capacity to assimilate it.* Farmers are well aware of the nutrient content of the fertilizer they apply, but may be less knowledgeable about nutrients in other inputs to their soils, such as crop residues or manure. Those who look to fertilizer for their crop's nutrient needs sometimes apply manure simply to get rid of it, without accounting for the additional nutrients that the material adds to the soil. The resulting nutrient overload leads to pollution of nearby water. But education can prevent much of this overapplication. The U.S. state of Maryland, for example, initiated a nutrient management program in 1989 to help farmers monitor and control flows of nutrients on their farms. In just seven years, the program has enrolled well over half of the state's cultivated croplands, and nutrient overapplications on them have been reduced. In 1996, program consultants recommended an average reduction of 15 pounds of nitrogen per acre.[85]

Once they internalize these principles, citizens and policymakers essentially achieve a major shift in thinking and in world view. Organic matter is no longer seen as disposable garbage, but as a soil-building natural resource. And nutrients are no longer viewed as wholly benign, to be scattered wantonly throughout the environment, but are understood to serve economies and ecosystems best when kept in balance. Such reacceptance of the ancient appreciation of organic material will be a large step in the direction of building sustainable cities and farms.

Notes

1. Grain to Rome from "History of North Africa," *Encyclopedia Britannica*, vol. 13 (macropaedia), 1976; environmental decline from Herbert Girardet, "Cities and the Biosphere," paper presented to the UNDP-Marmaris Roundtable, Cities for People in a Globalizing World, 19-21 April 1996.

2. Drinking water from Organisation for Economic Co-operation and Development (OECD), "Towards Sustainable Agricultural Production: Cleaner Technologies" (Paris, 1994); species diversity from David Wedin and David Tilman, "Influence of Nitrogen Loading and Species Composition on the Carbon Balance of Grasslands," *Science*, 6 December 1996; quality of organic matter from Harry A.J. Hoitink, et al., "Making Compost to Suppress Plant Disease," *Biocycle*, April 1997; Gulf of Mexico from Jonathan Tolman, "Poisonous Runoff from Farm Subsidies," *Wall Street Journal*, 8 September 1995.

3. Flush toilets from Peter Gleick, "Basic Water Requirements for Human Activities: Meeting Basic Needs," *Water International*, June 1996.

4. Share of nutrients in municipal waste is a Worldwatch calculation based on data from OECD, *OECD Environmental Data: Compendium 1995* (Paris: 1995), on U.S. waste data from Environmental Protection Agency (EPA) "Characterization of Municipal Solid Waste in the United States: 1995 Update," Executive Summary (Washington, DC: March 1996), on nutrient value of municipal waste from Xin-Tao He, Terry J. Logan, and Samuel J. Traina, "Physical and Chemical Characteristics of Selected U.S. Municipal Solid Waste Composts," *Journal of Environmental Quality*, May–June 1995, and on fertilizer use from Food and Agricultural Organization (FAO), FAO web site http://www.fao.org. Countries selected were those for which complete data was available. Share of nutrients in human waste does not include the 33 percent of sludge produced in OECD countries that is already applied to land. Share is a Worldwatch calculation based on nutrient value of human waste from E. Witter and J.M. Lopez-Real, "The Potential of Sewage Sludge and Composting in a Nitrogen Recycling Strategy for Agriculture," *Biological Agriculture and Horticulture*, 5, 1987; population from U.S. Agency for International Development (USAID) and U.S. Department of Commerce, *World Population Profile 1996*, and on fertilizer use from FAO, op.cit. this note.

5. OECD, op. cit. note 4. Share is an average of member states' reporting rates for the early 1990s. If paper is included in the organic total, the organic share rises to nearly two thirds. Composting toilets from Robert Goodland and Abby Rockefeller, "What Is Environmental Sustainability in Sanitation?" *UNEP-IETC Newsletter*, Summer 1996.

6. Teresa Glover, "Livestock Manure: Foe or Fertilizer?" *Agricultural Outlook*, June 1996.

7. China from A.E. Johnston, "The Efficient Use of Plant Nutrients in Agriculture," (Paris: International Fertilizer Industry Association, 1995); 23 states from Nora Goldstein, "The State of Garbage in America," *Biocycle*, April 1997; United States from National Research Council, *Use of Reclaimed Water and Sludge in Food Crop Production* (Washington, DC: National Academy Press, 1996); Europe from Peter Matthews, ed., *A Global Atlas of Wastewater Sludge and Biosolids Use and Disposal* (London: International Association on Water Quality, 1996) and Peter Matthews, Director of Innovation, Anglian Water, Cambridgeshire, U.K., letter to author, 24 June 1997.

8. Nitrogen fixation from Peter M. Vitousek et al., "Human Alteration of the Global Nitrogen Cycle: Causes and Consequences," *Ecological Issues*, February 1997; fossil-fuel burning and cultivation of nitrogen-fixing crops are the other human sources of nitrogen fixation. Still other human activities—the burning of forests, wood fuel, and grasslands; draining of wetlands; and clearing of land for crops—release trapped nitrogen that was already fixed. Figure 1 from Vitousek et al., op. cit. this note, and Ann P. Kinzig and Robert H. Socolow, "Human Impacts on the Nitrogen Cycle," *Physics Today*, November 1994. Fertilizer application from Lester R. Brown, "Fertilizer Use Rising Again," *Vital Signs 1997: The Environmental Trends That Are Shaping Our Future* (New York: W.W. Norton and Company, 1997).

9. Wedin and Tilman, op. cit. note. 2; northern Europe from Vitousek, op. cit. note 8, and C. Mlot, "Tallying Nitrogen's Increasing Impact," *Science News*, 15 February 1997.

10. Wedin and Tilman, op. cit. note 2.

11. L. Drinkwater, P. Wagoner, M. Sarrantino, and S. E. Peters, "Net primary productivity, nitrogen balance and carbon sequestration in organic and conventional maize/soybean cropping systems," submitted for publication to *Ecological Applications*, and Laurie Drinkwater, Rodale Institute Research Center, Kutztown, PA, letter to author, 12 June 1997. Like the conventionally fertilized field, the manure-fed fields received more nitrogen than was taken up by crops. But unlike the conventional field, which had a high rate of leaching, the manure-fed land was effective at storing nitrogen for later use by crops.

12. Tolman, op. cit. note 2.

13. OECD, op. cit. note 2; India and Brazil from "Comprehensive Assessment of the Freshwater Resources of the World," report to the United Nations, UN Commission on Sustainable Development web site <http://www.un.org/dpcsd/dsd/freshwat.htm>, viewed 14 March 1997.

14. Content of organic matter from Nyle C. Brady and Ray R. Weil, *The Nature and Properties of Soils*, 11th edition (Upper Saddle River, NJ: Prentice Hall, 1996).

15. African fertilizer use from Amitava Roy, "Nutrient Inputs as Critical Variables in the Long-Term Projections for Sustainable Global Food Security," unpublished paper (Muscle Shoals, AL: International Fertilizer Development Center, undated).

16. Landfill closures from Goldstein, op. cit. note 7; capacity increase from Council on Environmental Quality, *Environmental Quality* (Washington, DC: U.S. Government Printing Office, 1997); Fresh Kills from Vivian S. Toy, "Bids for Exporting Trash Are Lower than Expected," *New York Times*, 3 March 1997; Institute for Local Self-Reliance, *Beyond 25 Percent: Materials Recovery Comes of Age* (Washington, DC, 1989).

17. OECD, op. cit. note 4. Share is an average of member states' reporting rates for the early 1990s. Including paper in the organic total boosts the organic share to nearly two thirds; leaching and methane from EPA, "Yard Waste Composting," Environmental Fact Sheet, (Washington, DC: Office of Solid Waste, January 1991); Fresh Kills from Nancy Reckler, "New Yorkers Near World's Largest Landfill Say City Dumps on Them," *Washington Post*, 7 August 1996.

18. Robert Steuteville, "The State of Garbage in America," *Biocycle*, April 1996; Alastair Guild, "Britain's landfill tax raises stakes for compost makers," *Financial Times*, 10 October 1996; "Tokyo Examines Fees for Collection of Garbage from Households by 1999," *International Environmental Reporter*, 5 February 1997; Paul Relis and Howard Levenson, "Using Urban Organics in Agriculture," *Biocycle*, April 1997.

19. John Briscoe and Mike Garn, "Financing Agenda 21: Freshwater," paper prepared for the United Nations Commission on Sustainable Development (Washington, DC: World Bank, February 1994).

20. Water stress from "Comprehensive Assessment," op. cit. note 13; flush toilets from Gleick, op. cit. note 3; cost from Briscoe and Garn, op. cit. note 19.

21. "Photosynthesis" in *Encyclopedia Britannica*, vol. 14 (macropedia), 1976.

22. Share of calories from Tim Dyson, *Population and Food: Global Trends and Future Prospects* (London: Routledge, 1996).

23. U.S. Department of Agriculture (USDA), *Production, Supply, and Distribution (PS&D)*, electronic database, Washington, DC, updated October 1996; Table 1 based on data in USDA, op. cit. this note.

24. USDA, op. cit. note 23.

25. USDA, op. cit. note 23.

26. Ten percent calculated from data in USDA, op. cit. note 23; Table 2 based on data in USDA, op. cit. note 23.

27. G.W. Cooke, "The Intercontinental Transfer of Plant Nutrients," in *Nutrient Balances and the Need for Potassium*, Proceedings of the 13th International Potash Institute Congress, August 1986, Reims, France (Basel, Switzerland: International Potash Institute, 1986); Table 3, op. cit. this note.

28. African flows from Cooke, op. cit. note 27; Roy, op. cit. note 15.

29. Overapplication based on data in USDA, op. cit. note 23; Table 4 based on data in USDA, op. cit. note 23.

30. Recycling rate of manure from Council for Agricultural Science and Technology (CAST) *Integrated Animal Waste Management*, Task Force Report No. 128 (Ames, IA, November 1996).

31. Asia from Johnston, op. cit. note 7; United States from National Research Council, op. cit. note 7; Europe from Matthews, ed., op. cit. note 7, and Matthews, op. cit. note 7.

32. Organic share is an average of member states' reporting rates for the early 1990s, and is calculated from data in OECD, op. cit. note 4; including paper in the organic total boosts the organic share to nearly two thirds. Developing countries from Centre de Cooperation Suisse pour la Technologie et le Management (SKAT), *Valorisation des dechets organiques dans les quartiers populaires des villes africaines* (St. Gallen, Switzerland, 1996); 11 percent is an average of member states' reporting rates for the early 1990s, and is based on data in OECD, op. cit. note 4.

33. Volatilization from Witter and Lopez-Real, op. cit. note 4; high-yielding varieties from Balu L. Bumb and Carlos A. Baanante, "The Role of Fertilizer in Sustaining Food Security and Protecting the Environment to 2020," Food, Agriculture, and the Environment Discussion Paper 17 (Washington, DC: International Food Policy Research Institute, September 1996).

34. Organic share is an average of member states' reporting rates for the early 1990s, and is calculated from data in OECD, op. cit. note 4; including paper in the organic total boosts the organic share to nearly two thirds; developing countries from *Valorisation*, op. cit. note 32; 11 percent is an average of member states' reporting rates for the early 1990s, and is based on data in OECD, op. cit. note 4; Table 5 calculated from data in OECD, op. cit. note 4.

35. Paper data from OECD, op. cit. note 4.

36. Brady and Weil, op. cit. note 14.

37. H.A.J. Hoitink, A.G. Stone, and D.Y. Han, "Suppression of Plant Diseases by Composts," accepted for publication in *HortScience*, 1997;

replacement of methyl bromide from William Quarles and Joel Grossman, "Alternatives to Methyl Bromide in Nurseries—Disease Suppressive Media," *IPM Practitioner*, August 1995.

38. For calculation of share of nutrients in municipal waste, see note 4. Estimate of nutrient overapplication based on Dale Lueck, "Policies to Reduce Nitrate Pollution in the European Community and Possible Effects on Livestock Production," Economic Research Service (Washington, DC: USDA, September 1993), which reports nitrogen overapplication in Europe to be 57 percent greater than crop needs; on U.S. grain data from USDA, op. cit. note 23, and fertilizer application rates from Economic Research Service, "Agricultural Resources," February 1993, which are used to calculate a fertilizer overapplicaton rate in the U.S. of 36 percent; Table 6 is Worldwatch calculation based on data from OECD, EPA, and FAO, op. cit. note. 4. Countries selected were those for which complete data was available.

39. Slow release from Brady and Weil, op. cit. note 14; India from Panneer Selvam, "A Review of Indian Experiences in Composting of Municipal Solid Wastes and a Case Study on Private Sector Participation," paper presented to the Conference on Recycling Waste for Agriculture: The Rural-Urban Connection, held at the World Bank, Washington, DC, 23-24 September 1996.

40. Toni Nelson, "Closing the Nutrient Loop," *World Watch*, November/December 1996.

41. Integrated Waste Management Board, "Agriculture in Partnership with San Jose," Final Report, (Sacramento, CA: Integrated Waste Management Board, April 1997); need for customizing from Francis R. Gouin, "Compost Use in the Horticultural Industries," *Green Industry Composting*, undated.

42. Composting facilities from Goldstein, op. cit. note 7; San Jose from Integrated Waste Management Board, "Agriculture in Partnership with San Jose," Final Report (Sacramento, CA, April 1997).

43. Dave Baldwin, Community Recycling and Resource Recovery, Inc., Lamont, CA, conversation with author, 21 February 1997.

44. Value of compost from Community Recycling and Resource Recovery, Inc., "Community Recycling Compost Typical Analysis," factsheet (Lamont, CA, undated).

45. "Institutions Save by Composting Food Residuals," *Biocycle*, January 1997.

46. Selvam, op. cit. note 39.

47. China from Alice B. Outwater, *Reuse of Sludge and Minor Wastewater Residuals* (Boca Raton, FL: Lewis Publishers, 1994). While sludge is already recycled with few problems in many countries, the risk lies in the lack of control over materials that enter sewers. This lack of control means that any batch of sludge could contain materials that are harmful to human or environmental health. Laura Orlando, Resource Institute for Low-Entropy Systems, Boston, MA, conversation with author, 18 June 1997.

48. Regional distinctions from World Resources Institute (WRI), *World Resources 1996–97* (New York: Oxford University Press, 1996); 10 percent from Witter and Lopez-Real, op. cit. note 4; arid regions from Carl R. Bartone, "International Perspective on Water Resources Management and Wastewater Reuse—Appropriate Technologies," *Water Science Technology*, 23, 1991.

49. Official encouragement from Peter Matthews, ed., op. cit. note 7; reuse rates: United States from National Research Council, op. cit. note 7; Europe from Matthews, ed., op. cit. note 7, and Matthews, op. cit. note 7; dumping sites from National Research Council, op. cit. note 7. Ocean dumping, once a common method of sewage disposal for some coastal cities, was outlawed in the United States in 1992, and will be illegal in Europe after 1998. See Cecil Lue-Hing, Peter Matthews, Juraj Namer, Nagaharu Okuno, and Ludovico Spinosa, "Sludge Management in Highly Urbanized Areas," in Matthews, ed., op. cit. note 7.

50. Assumes that 33 percent of sludge produced in OECD countries is already land applied. For calculation of share of nutrients in human waste, see note 4. For an explanation of the estimate of nutrient overapplication, see note 38; Table 7 is Worldwatch calculation based on data as follows: fertilizer use from FAO, op. cit. note 4; nutrient value of human waste from Witter and Lopez-Real, op. cit. note 4; population from USAID and U.S. Department of Commerce, op. cit. note 4.

51. Bartone, op. cit. note 48.

52. Israeli reuse from Sandra Postel, "Dividing the Waters: Food Security, Ecosystem Health, and the New Politics of Scarcity," Worldwatch Paper 132 (Washington, DC: Worldwatch Institute, September 1996); heavy metal levels from Yoram Avnimelech, "Irrigation with Sewage Effluents: The Israeli Experience," *Environmental Science and Technology*, 27, no. 7, 1993.

53. Outbreaks from Hillel I. Shuval, "Wastewater Irrigation in Developing Countries: Health Effects and Technical Solutions," Summary of World Bank Technical Paper Number 51 (Washington, DC: World Bank, 1990); Mexico from Duncan Mara and Sandy Cairncross, *Guidelines for the Safe Use of Wastewater and Excreta in Agriculture and Aquaculture* (Geneva: World Health Organization, 1989).

54. Substances from Laura Orlando, "The Sewage Scam: Should Sludge Fertilize Your Vegetables?" *Dollars and Sense*, May/June 1997; persistence of metals in soils from "Land Application of Sewage Sludge," excerpt from

"Cornell Recommends," in press, August 1996.

55. Standards from Orlando, op. cit. note 54; testing from Mark Lang, Carolyn E. Jenkins, and W. Dale Albert, "USA: Northeastern States," in Peter Matthews, ed., op. cit. note 7.

56. Ponds from Bartone, op. cit. note 48; pathogen kill from D.D. Mara, G.P. Alabaster, H.W. Pearson, and S.W. Mills, "Waste Stabilization Ponds: A Design Manual for Eastern Africa" (Leeds, U.K.: Lagoon Technology International, 1992). Other sources list the rate of pathogen kill in conventional plants as only 90-95 percent. See Shuval, op. cit. note 53.

57. Peter Edwards, *Reuse of Human Wastes in Aquaculture,* Water and Sanitation Report No. 2, UNDP-World Bank Water and Sanitation Program, 1992. Multiple uses from Dhrubajyoti Ghosh, "Wastewater-Fed Aquaculture in the Wetlands of Calcutta—an Overview," in P. Edwards and R.S.V. Pullin, *Wastewater-Fed Aquaculture,* Proceedings of the International Seminar on Wastewater Reclamation and Reuse for Aquaculture, Calcutta, India, 6–9 December 1988.

58. Living Technologies, "What Is a Living Machine?" (factsheet) (Burlington, VT, 1997).

59. Pond area from Shuval, op. cit. note 53; Calcutta from Carl R. Bartone, op cit. note 48.

60. SIRDO's full name is Sistema Integral de Reciclamiento de Desechos Organicos, or Integral System for Recycling Organic Waste.

61. Grupo de Tecnologia Alternativa (GTA), "The SIRDO from Mexico, 1979-1992" (Mexico City: Grupo de Tecnologia Alternativa, undated); Josefina Mena Abraham, Grupo de Tecnologia Alternativa, Mexico City, e-mail to author, 16 June 1997, and Sidonie Chiapetta, National Wildlife Federation (NWF), e-mail to author, 18 June 1997.

62. Sidonie Chiapetta, NWF, conversation with author, 17 June 1997; Tres Marias from Josefina Mena Abraham, Grupo de Tecnologia Alternativa, Mexico City, letter to author, 20 February 1996.

63. Sidonie Chiapetta, "Costs and Benefits of SIRDO Technology," information sheet, (Washington, DC: NWF, September 1996). The NWF analysis works out to $33 per household per year (assuming 7-8 persons per household). The GTA estimates net household revenues of $30-60 per year, Abraham, op. cit. note 61.

64. Chiapetta, op. cit. note 63.

65. J. Paul Henderson, "Anaerobic Digestion in Rural China," *Biocycle,* January 1997.

66. Carol Steinfeld, "Compost Toilets Reconsidered," *Biocycle*, March 1997.

67. Orlando, op. cit. note 47.

68. Grain output and exports from USDA, op. cit. note 23; trade and production from FAO, "Characteristics of Agricultural Trade," in *World Food Summit, Technical Background Documents 12-15, Volume 3* (Rome, 1996).

69. Angela Paxton: "The Food Miles Report: The Dangers of Long-Distance Food Transport" (London: SAFE Alliance, September 1994).

70. Nitrate levels from Dale J. Leuck, "Policies to Reduce Nitrate Pollution in the European Community and Possible Effects on Livestock Production," Economic Research Service (Washington, DC: USDA, September 1993).

71. Corn from USDA, op. cit. note 23; Taiwan pollution from USDA, "U.S. Grain Producers Have Big Steak in Taiwan's Market," *Grain: World Markets and Trade*, June 1997.

72. Glover, op. cit. note 6.

73. Table 8 based on Glover, op. cit. note 6.

74. Missouri farm from Mark Schultz, Land Stewardship Project, Minneapolis, MN, discussion with author, March, 1997; pollution from Glover, op. cit. note 6.

75. Brady and Weil, op. cit. note 14.

76. SKAT, op. cit. note 32.

77. Selvam, op. cit. note 39.

78. E.I. Stentiford, J.T. Pereira Neto, and D.D. Mara, *Low cost composting*, Research Monograph No. 4 (Leeds, U.K.: Dept. of Civil Engineering, University of Leeds, 1996).

79. Inge Lardinois and Arnold van de Klundert, "Recycling Urban Organics in Asia and Africa," *Biocycle*, June 1994.

80. Steuteville, op. cit. note 18; Guild, op. cit. note 18.

81. Cut in flows from Paul Vossen and Ellen Rilla, "Trained Home Composters Reduce Solid Waste by 18%," *California Agriculture*, September-October 1996; costs from Ellen Rilla, "CE Offices Facilitate Community Composting Efforts," *California Agriculture*, September-October 1996.

82. Funding limitations and composting toilets from Peter H. Gleick, ed., *Water in Crisis: A Guide to the World's Fresh Water Resources* (New York:

Oxford University Press, 1993). Note that a much higher cost differential—28 times—is given in Briscoe and Garn, op. cit. note 14; Sweden from Goodland and Rockefeller, op. cit. note 5.

83. Bob Scrowcroft, Organic Farming Research Foundation, Santa Cruz, CA, conversation with author, 25 June 1997.

84. Emily Green and Jim Kleinschmidt, "Nutrient Management Yardsticks" information sheet (Minneapolis, MN: Institute for Agriculture and Trade Policy, 1996).

85. Cooperative Extension Service, "Maryland Nutrient Management Program Annual Report, 1996" (College Park, MD: Maryland Department of Agriculture, 1996).

PUBLICATION ORDER FORM

_____ *State of the World:* **$13.95**
The annual book used by journalists, activists, scholars, and policymakers worldwide to get a clear picture of the environmental problems we face.

_____ *Vital Signs:* **$12.00**
The book of trends that are shaping our future in easy to read graph and table format, with a brief commentary on each trend.

_____ **Subscription to WORLD WATCH magazine: $20.00 (international airmail $35.00)**
Stay abreast of global environmental trends and issues with our award-winning, eminently readable bimonthly magazine.

_____ **Worldwatch Library: $30.00 (international subscribers $45)**
Receive *State of the World* and all Worldwatch Papers as they are released during the calendar year.

_____ **Worldwatch Database Disk Subscription: $89.00**
Contains global agricultural, energy, economic, environmental, social, and military indicators from all current Worldwatch publications including this Paper. Includes a mid-year update, and *Vital Signs* and *State of the World* as they are published. Can be used with Lotus 1-2-3, Quattro Pro, Excel, SuperCalc and many other spreadsheets.   **Check one:** _____ **IBM-compatible or** _____ **Macintosh**

_____ **Worldwatch Papers—See complete list on following page**
Single copy: $5.00 • 2–5: $4.00 ea. • 6–20: $3.00 ea. • 21 or more: $2.00 ea. (Call Director of Communication, at (202) 452-1999, for discounts on larger orders.)

$4.00 Shipping and Handling *($8.00 outside North America)*

_____ **TOTAL**

Make check payable to Worldwatch Institute
1776 Massachusetts Ave., NW, Washington, DC 20036-1904 USA

Enclosed is my check or purchase order for U.S. $_____

☐ AMEX ☐ VISA ☐ MasterCard _____
                                   Card Number   Expiration Date

_____
name   **daytime phone #**

_____
address

_____
city   **state**   **zip/country**

phone: (202) 452-1999   fax: (202) 296-7365   e-mail: wwpub@worldwatch.org
website: www.worldwatch.org

☐ **Send me a brochure of all Worldwatch publications.**

Worldwatch Papers

No. of Copies

_____**Total copies (transfer number to order form on previous page)**

PUBLICATION ORDER FORM

_____ *State of the World:* **$13.95**
The annual book used by journalists, activists, scholars, and policymakers worldwide to get a clear picture of the environmental problems we face.

_____ *Vital Signs:* **$12.00**
The book of trends that are shaping our future in easy to read graph and table format, with a brief commentary on each trend.

_____ **Subscription to WORLD WATCH magazine: $20.00 (international airmail $35.00)**
Stay abreast of global environmental trends and issues with our award-winning, eminently readable bimonthly magazine.

_____ **Worldwatch Library: $30.00 (international subscribers $45)**
Receive *State of the World* and all Worldwatch Papers as they are released during the calendar year.

_____ **Worldwatch Database Disk Subscription: $89.00**
Contains global agricultural, energy, economic, environmental, social, and military indicators from all current Worldwatch publications including this Paper. Includes a mid-year update, and *Vital Signs* and *State of the World* as they are published. Can be used with Lotus 1-2-3, Quattro Pro, Excel, SuperCalc and many other spreadsheets. **Check one:** _____ **IBM-compatible or** _____ **Macintosh**

_____ **Worldwatch Papers—See complete list on following page**
Single copy: $5.00 • 2–5: $4.00 ea. • 6–20: $3.00 ea. • 21 or more: $2.00 ea. (Call Director of Communication, at (202) 452-1999, for discounts on larger orders.)

$4.00 Shipping and Handling *($8.00 outside North America)*

_____ **TOTAL**

Make check payable to Worldwatch Institute
1776 Massachusetts Ave., NW, Washington, DC 20036-1904 USA

Enclosed is my check or purchase order for U.S. $_____

☐ AMEX   ☐ VISA ☐ MasterCard _____
                                                       Card Number   Expiration Date

_____
name   **daytime phone #**

_____
address

_____
city   **state**   **zip/country**

phone: (202) 452-1999   fax: (202) 296-7365   e-mail: wwpub@worldwatch.org
website: www.worldwatch.org

☐ **Send me a brochure of all Worldwatch publications.**

Worldwatch Papers

No. of Copies

_____**Total copies (transfer number to order form on previous page)**